T0306359

Leveraging Complexity in Great-Power Competition and Warfare

Volume II, Technical Details for a Complex Adaptive Systems Lens

SHERRILL LINGEL, MATTHEW SARGENT, TIMOTHY R. GULDEN, TIM MCDONALD, PAROUSIA ROCKSTROH

Prepared for the Department of the Air Force
Approved for public release; distribution unlimited

For more information on this publication, visit **www.rand.org/t/RRA589-2**.

About RAND

The RAND Corporation is a research organization that develops solutions to public policy challenges to help make communities throughout the world safer and more secure, healthier and more prosperous. RAND is nonprofit, nonpartisan, and committed to the public interest. To learn more about RAND, visit www.rand.org.

Research Integrity

Our mission to help improve policy and decisionmaking through research and analysis is enabled through our core values of quality and objectivity and our unwavering commitment to the highest level of integrity and ethical behavior. To help ensure our research and analysis are rigorous, objective, and nonpartisan, we subject our research publications to a robust and exacting quality-assurance process; avoid both the appearance and reality of financial and other conflicts of interest through staff training, project screening, and a policy of mandatory disclosure; and pursue transparency in our research engagements through our commitment to the open publication of our research findings and recommendations, disclosure of the source of funding of published research, and policies to ensure intellectual independence. For more information, visit www.rand.org/about/principles.

RAND's publications do not necessarily reflect the opinions of its research clients and sponsors.

Preface

Modern warfare is characterized by a complex environment of operations spanning multiple domains leveraged to achieve advantage against the adversary in both strategic competition and armed conflict. However, understanding complexity in warfare, often referred to as the *art of war*, is rarely framed in a structured way. This report seeks to provide an initial examination of how complex adaptive systems thinking can frame opportunities and challenges of complexity in warfare. This report is the second of a two-volume report. It provides the technical details in support of the first volume, *Leveraging Complexity in Great-Power Competition and Warfare: Volume I, An Initial Exploration of How Complex Adaptive Systems Thinking Can Frame Opportunities and Challenges.* Three topics are covered in this volume: a literature review of complex adaptive systems research, a complex adaptive systems lens user manual, and a Markov chain formulation of the adversary decision process. This report should be of interest to warfighting planners, wargamers, complex adaptive systems researchers, and the military science and technology community.

The research reported here was commissioned by the Air Force Research Laboratory executive director, Jack Blackhurst, and conducted within the Force Modernization and Employment Program of RAND Project AIR FORCE as part of a fiscal year 2019 add-on project Complexity Imposition.

RAND Project AIR FORCE

RAND Project AIR FORCE (PAF), a division of the RAND Corporation, is the Department of the Air Force's (DAF's) federally funded research and development center for studies and analyses, supporting both the United States Air Force and the United States Space Force. PAF provides DAF with independent analyses of policy alternatives affecting the development, employment, combat readiness, and support of current and future air, space, and cyber forces. Research is conducted in four programs: Strategy and Doctrine; Force Modernization and Employment; Manpower, Personnel, and Training; and Resource Management. The research reported here was prepared under contract FA7014-16-D-1000.

Additional information about PAF is available on our website: www.rand.org/paf/

This report documents work originally shared with DAF in March 2020. The draft report, also issued in March 2020, was reviewed by formal peer reviewers and DAF subject-matter experts.

Contents

Figures

Tables

Summary

Issues

A key concern for the U.S. Air Force is the ability to leverage multidomain operations (MDOs) to U.S. advantage both in competition and in warfighting. Multidomain actions are viewed as imposing complexity on the adversary's decision process. But what does *complexity* mean in an Air Force context?

- Currently, we lack understanding of how to impose or exploit complexity to maximize operational effects.
- An informed understanding of the nature and value of complexity as a weapon is needed—that is, how complexity can be used as a method of attack. How can complexity be employed in an operational setting?
- Science and technology investments are not currently aligned to quantify complexity, measure its operational effects, and determine how to impose complexity and thus shape adversary behavior. How can research on the nature of complexity be used to understand what science and technology efforts might deliver in terms of complexity-informed capabilities?

Approach

In this technical volume, we provide context and supporting material to the framework in *Leveraging Complexity in Great-Power Competition and Warfare:* Volume I, *An Initial Exploration of How Complex Adaptive Systems Thinking Can Frame Opportunities and Challenges*.

- Chapter 2 provides a more-thorough discussion of complex adaptive system (CAS), informed by a survey of literature. Although the concept of complexity is widely represented among academic disciplines and professional fields, there is no unified definition. Given this, and attempting to make the concept of complexity tractable for military decisionmakers, we sought to identify a set of essential characteristics of complex systems that are common across multiple fields. We then used those characteristics to identify areas where a CAS could be exploited to U.S. advantage against an adversary or, conversely, to expose U.S. vulnerabilities. Together, these properties and avenues for imposition on a given system create a CAS *lens* that enables an analyst or decisionmaker to identify whether, how, and to what degree a system or situation is complex and how that complexity might be exploited.
- Chapter 3 expands on this lens, providing a rubric for assessing Air Force Research Laboratory investments in science and technology and operational capabilities and a "user manual" on how to use the rubric. Viewing operations or force planning through the CAS lens suggests various actions that take advantage of the complex adaptive nature of systems. The tool presented in this chapter helps the user recognize the adversary's

complexity challenges (if any) and identify ways to exploit them to U.S. advantage. The tool illuminates the adversary's organization, adaptive mechanisms, and emergent properties (if any). It then indicates potential vulnerabilities to look for and exploit.

- In Chapter 4, we expand on the adversary decision process presented in Chapter 3 of Volume I and develop the supporting mathematical material, providing further detail on Markov chains, transition matrixes, and modeling. We use properties of graphs and Markov chains to study the structure of the decision processes presented in Volume I. More specifically, we demonstrate how, by viewing a Markov chain as a stochastic process in a directed graph and the associated transition matrix, an analyst can study the structure of the aforementioned decision processes at two levels of resolution. We show that, with sufficient data, an analyst can use the Markov chain framework to calculate the outcome of the respective decision processes. However, if data are insufficient to perform these calculations (which is often the case), an analyst can still use the underlying graph structure of the Markov chain to understand the structure of the decision space. These two levels of fidelity show the utility of the Markov chain framework.

Acknowledgments

We would like to first thank the Air Force Research Laboratory Information Directorate sponsors, Lee Seversky and David Myers, thought leaders on command and control and complexity, who were so engaged with the RAND team on this project. This work would not exist without their intellectual curiosity and patience. We benefited greatly from the contributions of our colleges, who provided initial feedback on complex adaptive systems and multidomain warfare in two workshops: Jim Chow, Paul Davis, Dave Frelinger, Tom Hamilton, Karl Mueller, Angela O'Mahony, Dave Ochmanek, Dave Shlapak, Michael Spirtas, and Rand Waltzman.

Abbreviations

AFRL	Air Force Research Laboratory
C2	command and control
CAS	complex adaptive system
ISR	intelligence, surveillance, and reconnaissance
S&T	science and technology

1. Introduction to Volume II

In this technical volume, we provide context and supporting material to the frameworks in *Leveraging Complexity in Great-Power Competition and Warfare*: Volume I, *An Initial Exploration of How Complex Adaptive Systems Thinking Can Frame Opportunities and Challenges*. Chapter 2 of this volume provides a more-thorough discussion of complex adaptive systems (CAS), informed by a survey of literature. Although the concept of complexity is widely represented among academic disciplines and professional fields, there is no unified definition. Given this, and attempting to make the concept of complexity tractable for military decisionmakers, we sought to identify a set of essential characteristics of complex systems that are common across multiple fields. We then used those characteristics to identify areas where CAS could be exploited to U.S. advantage against an adversary or, conversely, to expose U.S. vulnerabilities. Together, these properties and avenues for imposition on a given system create a CAS *lens* that enables an analyst or decisionmaker to identify whether, how, and to what degree a system or situation is complex and how that complexity might be exploited.

Chapter 3 expands on this lens, providing a rubric for assessing Air Force Research Laboratory (AFRL) investments in science and technology (S&T) and operational capabilities and a "user manual" on how to use the rubric. Viewing operations or force planning through the CAS lens suggests various actions that take advantage of the complex adaptive nature of systems. The tool presented in this chapter helps the user recognize the adversary's complexity challenges (if any) and identify ways to exploit them to U.S. advantage. The tool illuminates the adversary's organization, adaptive mechanisms, and emergent properties (if any). It then indicates potential vulnerabilities to look for and exploit.

In Chapter 4, we expand on the adversary decision process presented in Chapter 3 of Volume I and develop the supporting mathematical material, providing further detail on Markov chains, transition matrixes, and modeling. We use properties of graphs and Markov chains to study the structure of the decision processes presented in Volume I. More specifically, we demonstrate how, by viewing a Markov chain as a stochastic process in a directed graph and the associated transition matrix, an analyst can study the structure of the aforementioned decision processes at two levels of resolution. We show that, with sufficient data, an analyst can use the Markov chain framework to calculate the outcome of the respective decision processes. However, if data are insufficient to perform these calculations (which is often the case), an analyst can still use the underlying graph structure of the Markov chain to understand the structure of the decision space. These two levels of fidelity show the utility of the Markov chain framework.

1

2. Complex Adaptive Systems

Complexity is central to the military experience. Leadership and personnel regularly work under conditions of uncertainty, coordinating their areas of responsibility with those of colleagues to achieve a shared objective. Over time, the development of heuristics has helped to make sense of these environments and guide analysis and action. Such concepts as deception, fog of war, and horns of multiple dilemmas are representations of complexity (Lawson, 2013).

Although complexity has been recognized since the earliest days of human society and warfare (Capra and Luisi, 2016), complexity has increasing salience in the modern era, driven by global social and technical trends, such as increased interconnectedness, diffusion of authority, and systemic uncertainty. Digital technologies are increasing the connectedness of societies and economies, changing how war is conducted. The prominence of gray-zone and hybrid warfare is indicative of the value of exploiting complex environments using such tools as deception, disinformation, operating below thresholds, and integrating multiple types of weapons or actions (Mazarr, 2016; Monaghan, 2019).

Technical understanding of the mathematics underlying modern complexity theory is not required to make its essential concepts comprehensible to decisionmakers. In this chapter, we seek to demystify the concept by outlining essential definitions and characteristics. We conducted a literature review, including writings by leading thinkers in complexity sciences and field-specific texts and papers, to identify a set of common characteristics that can be employed to recognize complex systems, analyze their behavior, and translate this to actions in warfare.

This chapter builds on the discussion of CAS in Volume I. A stronger appreciation of CAS is helpful not because it reveals new truths about the nature and characteristics of great-power competition and warfare; indeed, military analysts and officers of all ranks will recognize much of what is discussed in this chapter. Instead, a firm grasp of CAS provides analysts and officers alike with a new and powerful conceptual lens, new analytical tools, and an associated vocabulary for describing and navigating many of the more-complex aspects of modern warfare to gain advantage and mitigate one's own vulnerabilities.

Simple, Complicated, Complex, and Complex Adaptive Systems

We begin with a general discussion of system characteristics. The building blocks of systems are individual components or *elements*, and their relationships are *structure*. Systems consist of nodes (made up of elements) that are connected through relationships in what can be visualized as a *network of nodes*. The behavior of systems can be *linear*, meaning that changes in output are roughly proportional to changes in input, or *nonlinear,* where the relationship between input and output may change radically depending on system state. The individuals or organizations that

compose the nodes within a system are called *agents*, and they may or may not be *adaptive*—
adaptive agents can change behavior in response to one another and to changing external
conditions, which they in turn influence (discussed further in Table 2.1) (Bar-Yam, 1997;
Holland, 2014; Rickles, 2007).

Table 2.1. Simple, Complicated, Complex, and Complex Adaptive Systems

System Characterization	Number of Elements	Nonlinear	Adaptive Agents	Emergence	Examples
Simple	Small	No	No	No	A self-contained weapon system with limited parts, such as a gun, or a routine command structure
Complicated	Larger	Yes	No	No	Larger mechanical system, such as a drone or fighter jet
Complex (with no animate agents)	Usually large	Yes	No	Yes (because of inanimate features)	A drone taking off because of interaction of ground speed and lift
Complex adaptive	Larger (with exceptions)	Yes	Yes	Yes (because of animate agents or artificial intelligence)	A piloted aircraft adapting altitude, speed, and mode in response to enemy actions

SOURCE: Adapted from Davis et al., 2020.

When components of a system are combined, the result can be equal to or different from the
sum of the individual components. For example, when weights are progressively added to a
scale, the weight is summative, as are the forces that occur in a mechanism. Alternatively, when
single drops of water combine, they create the characteristic of wetness. This new property
arising from the combination of elements is called *emergence*—a central concept in complex
systems (Holland, 1999). We will discuss its different forms here.

Local emergence occurs within a limited scope without affecting the wider system, such as
when a small number of people gather in a town square and a small protest erupts. *Global
emergence* affects the wider environment, such as when crowd behaviors change public
sentiment. *Emergent complexity* refers to increasing complexity at a higher level, such as
significant physical or social effects that occur as a result of rising global temperatures.
Emergent simplicity, on the other hand, means a phenomenon that becomes simpler at higher
levels, such as moving from nuanced interpersonal relations to seemingly steady behavior among
large groups. Emergence is a form of *phase transition*, which occurs when a system moves from
one state to another, affecting the rules and relationships that govern the system. Protest to riot,
peace to war, or supersonic to hypersonic speed are examples of phase transitions.

An additional critical distinction among types of emergence is that between *behavioral
emergence* and *structural emergence* (Choate, 2017; Petrovich, 1999). A very basic example of
behavioral emergence would be the onset of a battle. If two armies are in a tense standoff, there
may be the occasional sniper shot or skirmish between patrols, but, at some point, the fire from
one side is met with fire from the other side. Then, a quick cycle of escalation sets in, leading to

a period of active battle. There is a qualitative difference between standoff (punctuated by the occasional incident) and a full-on battle (with rapid attacks and counterattacks). The battle may be started strategically by the senior commander on one side or the other, or it may gain momentum as the result of a multitude of decisions made by officers and enlisted personnel alike or, in the vocabulary introduced earlier, by agents at lower levels, each of whom are reacting to what they see or think they see.

Structural emergence involves the transfiguration of organizational units within a system. For example, air operations were originally thought of as an extension of ground operations, so early military aircraft were organized as part of the Army under the Army Air Corps. However, the differences between air and ground forces eventually grew stark enough to justify the creation of a new organization separate from the Army—the U.S. Air Force. A discussion of such differences, which range from differences in training to differences in organizational culture, is beyond the scope of this study. For an in-depth examination of current differences among the U.S. services, see Zimmerman et al. (2019). Air-related operations and concerns emerged as a structural cluster requiring a corresponding management structure.

Different bodies of literature commonly distinguish between complicated and complex systems, where a complicated system, such as a machine, can be understood by breaking it into its component parts and observing how each part relates to the others in the system (a process known as *reduction*). A complex system, on the other hand, has a higher degree of interconnection and multicausality (Nason, 2017), as explained later in further detail.

To expand on this idea, Davis et al. (2020) characterize four types of systems: simple, complicated, complex, and complex adaptive. These types of systems are outlined in Table 2.1.

In this description, *simple* systems have a small number of elements that interact in easily understood ways. These could include a self-contained mechanical weapon system with few parts, such as a handgun or rifle. It could also include routine actions among a command structure.

Complicated systems are typically larger with more elements and may have very sophisticated behaviors. Such systems can be understood by breaking them into parts and analyzing how each relates to the other in a direct and summative or linear fashion. A drone or fighter jet is a complicated machine. The components of a complicated system are inanimate and so do not adapt over time, and the system does not exhibit emergent behaviors.

Something that is *complex* has many parts that are interconnected and interdependent and may create emergent behaviors, such as a drone taking off because of the interaction of ground speed, physical structure, and lift. The system components are inanimate and do not adapt. It is not always clear what causes behaviors, and causes are almost always multisource. The system can demonstrate emergence, such as through a chemical reaction (M. Mitchell, 2009; S. Mitchell 2009).

Some complex systems can be further described as *CAS*. These include animate agents—people, teams, organizations—that account for their adaptive behavior. The systems change over time as the agents that constitute the systems frequently adjust their behavior to try to achieve their goals. Examples include horizontal and vertical teams within armed forces, individuals and organizations connected by cyber networks, and use of weapon systems (Davis et al., 2020; Miller and Page, 2007). For example, a piloted aircraft adapting altitude, speed, and mode (e.g., in initiating an attack) as a function of circumstances (e.g., enemy behavior) is both complex and adaptive. Artificial intelligence agents may substitute for animate agents (e.g., an artificial intelligence–piloted drone may be substituted for a human-piloted aircraft).

The typology articulated by Davis et al. (2020) provides a useful initial conceptual grounding of different kinds of systems and the distinctive aspects of CAS. As we look deeper into the bodies of literatures on complex systems and complexity in warfare, additional characteristics emerge.

Complexity in Warfare

In warfare, the people, institutions, equipment, terrain, and other relevant elements of conflict combine to form a CAS (Davis, 2006). The behavior on either side depends on the behavior of the adversary, as well as the details of one side's own dynamic position. Small changes can have large effects, as is captured in the adage about the loss of a horseshoe nail leading to the loss of a war.[1]

Complexity permeates conventional, asymmetric, hybrid, and gray-zone forms of conflict. Conventional warfare includes operations at multiple levels, with combinations of units and weapons being dependent on and interacting with each other. Asymmetric forms of conflict, such as terrorism and insurgency, exploit incomplete information and uncertainty and seek to find points of high leverage. Hybrid campaigns represent a combination of military operations with asymmetric techniques, such as the 2006 Lebanon War in which the Israeli Defense Forces fought against a much smaller paramilitary Hezbollah force (Monaghan, 2019).

Gray-zone conflict, in particular, is one of the clearest manifestations of complexity. It employs economic, political, and information warfare to achieve political ends (Mazarr, 2016). Russia's campaigns in Eastern Europe are examples of the use of mostly nonmilitary means and staying below thresholds to avoid provoking escalatory responses (Mazzocchi, 2008). By adopting this approach, the Russian forces demonstrated that they were "willing to edge gradually toward their objectives rather than making an all-out grab" (Mazarr, 2015).

Many existing and historical concepts are examples of complexity. *Influence operations* are defined in Air Force doctrine as efforts that

> employ capabilities to affect behaviors, protect operations, communicate
> commander's intent, and project accurate information to achieve desired effects

---

[1] This concept of small changes having large consequences is generally referred to as "sensitive reliance on initial conditions" and is a hallmark of complex systems. This characteristic makes the behavior of complex systems notoriously difficult to forecast, even when the workings and connections of their parts are well understood.

across the cognitive domain. These effects should result in differing behavior or a change in the adversary's decision cycle, which aligns with the commander's objectives. (Air Force Doctrine Document 3-13, 2011)

Deception, or *military deception*,

> is defined as those actions executed to deliberately mislead adversary decision makers as to friendly military capabilities, intentions, and operations, thereby causing the adversary to take specific actions (or inactions) that will contribute to the accomplishment of the friendly mission. (Joint Publication 3-13.4, 2006)

Fog of war is generally the uncertainty about enemy, environment, friendly forces, and laws of war (Von Clausewitz, 1989; Setear, 1989), and the *multiple horns of dilemmas* draws from a Greek logical dilemma in which someone is faced with a decision among multiple poor choices. Former Chief of Staff of the Air Force Gen David Goldfein further described multidomain warfare as using "dominance in one domain or many, blending a few capabilities or many, to produce multiple dilemmas for the adversary in a way that will overwhelm them" (Pope, 2019).

In recent decades, there has been expansion in capabilities for understanding and managing challenges posed by CAS. These capabilities range from analytic tools, such as network mapping and modeling, to methodologies for structuring decisionmaking in conditions of uncertainty.

Wargaming enables participants to work through events with human decisionmaking, leading to identification of patterns of behavior or unusual behaviors that may not have occurred without the game (Henry, Berner, and Shlapak, 2017). *Simulation* can run through multiple permutations of an event and help train operators, particularly when confronted with surprising, unexpected events or when connected with others and acting and reacting (Siegfried, 2014). *Adaptive planning* methodology enables planners to anticipate the unexpected and change course as new information becomes available. Innovative modeling approaches, such as *agent-based modeling*, provide ways to explore how settings can evolve when actions are taken and participants in a system (the agents) change their behavior in response to evolving conditions and the actions of others (Bosse, Sharpanskykh, and Treuer, 2010).

Complex Adaptive System Concept Map

Although we just offered a framework to differentiate complex systems and CAS, to make the concept of complexity tractable for decisionmakers, we sought to further delineate the characteristics commonly associated with complex systems. This clarity would enable an analyst or decisionmaker to be more precise about whether and how a situation or system is complex. Furthermore, there are not strict distinctions among simple, complicated, complex, and CAS, but instead systems can be understood as more or less complex.

We conducted a review of complex systems literature in a broad range of disciplines, including complexity sciences, computer science, philosophy, and the natural, physical, and social sciences. In the course of this review, we identified a number of characteristics that are consistently related to CAS. We then trimmed the list of characteristics to reduce concept overlap and then organized it into a CAS concept map suited for Air Force application. Figures 2.1 and 2.2 show the concept maps.

We organize the characteristics into two major categories: properties of CAS and avenues for exploiting complexity. In the sections that follow, we unpack each of the properties and actions in the concept map, describing them abstractly and pointing out how they might be relevant to Air Force concerns. This concept map provides the basis for the proposed CAS lens discussed later in this report and presented in Volume I.

Properties of Complex Adaptive Systems

CAS have characteristic properties, as outlined in Figure 2.1. This section is designed to provide a minimal working description of the properties of CAS. Conceiving of the system in these terms is the first step in viewing it through the CAS lens. We have kept this conception extremely simple to allow for intuitive application.

Figure 2.1. Properties of Complex Adaptive Systems

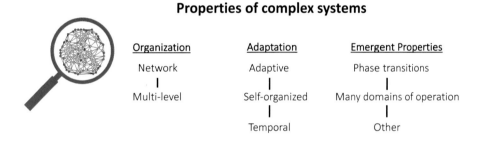

Organization

CAS are often organized as *networks* (Estrada, 2011) and tend to organize as a hierarchy of *nested nodes*. To be *nested* means that systems and networks are placed within larger systems and networks. The *nodes* are centralized authority within an otherwise diffuse system. In a social system (such as a conflict), these nodes in the network may be individual people, organizations, or even pieces of equipment. The links between nodes are called *edges*, and they consist of relationships of various sorts, referred to in network theory as *structure*. Depending on the situation, these edges could be command relationships, functional working relationships, friendships, channels of communication, or a host of other aspects of functional, professional, physical, or social connection.

Complex systems exist at *multiple levels*, such as by rank, personnel, or operations. Different levels of a system have different characteristics, behaviors, and rules, and higher-level systems assume the rules and behaviors of those beneath them but may also have their own characteristics (Mazzocchi, 2008; National Academies of Sciences, Engineering, and Medicine, 2019). For example, the nature of work and relationships among personnel of the same level performing a task is different from the relationships between them and higher-ranking or junior-ranking personnel.

The network connections associated with CAS tend to exhibit characteristic patterns. Militaries have a hierarchical command structure but also have horizontal connections at lower ranks between soldiers with interdependent responsibilities. CAS networks tend to have a property referred to by Herbert Simon as "near decomposability" (Simon, 1996), under which very large systems can be largely understood in terms of interacting subsystems.[2] The complex system can be *decomposed* into subsystems, and the functioning of those subsystems and the relationships among them can still be understood with reasonable, if imperfect, fidelity without getting into the details of each subsystem.

The human body, for example, comprises a very large number of cells, each of which plays a role in maintaining the health and integrity of the body. However, most of these cells are organized into tissues, and then the tissues into organs (and those that are not organized into tissues, such as red and white blood cells, travel throughout the blood and lymphatic systems performing functions specific to their subtypes). The organs can be understood in terms of their higher-level functions. A general understanding of what the heart does in the body does not require analysis of each of its constituent cells or even of its individual parts (e.g., chambers, valves, arteries, nerves). All of this can be abstracted away, and the heart can be thought of as a pump with a given capacity and endurance: It can be treated, for many purposes, as a "black box" that simply pumps blood through the body. The heart interacts with other organs, particularly the lungs, muscles, and brain, but a detailed cellular model is not needed to understand these interactions with a fair degree of fidelity for many purposes.

Following this conception, we can decompose a conflict CAS into three major components, which we will label Blue (the United States and its allied forces), Red (adversary forces), and other parties, such as civilian populations in the conflict area. Each of these components has many more connections among the nodes within it than it has to nodes in the other subsystems. Each of these major components can be further decomposed into military services (e.g., Army, Navy, Air Force) and then into increasingly granular functional and command structures to a level that is appropriate for understanding whatever the question at hand might be.

We can speak meaningfully of Blue response to Red, even though both Blue and Red comprise complex arrangements of people, doctrine, communications, and equipment, among other elements. Similarly, we can talk about a headquarters unit or a flight crew without understanding the functions of each of the people who compose these units. Blue, Red, headquarters, and flight crew are all examples of functional subsystems within the context of a larger complex system.

---

[2] The reductionist idea that a system that can be divided into pieces and behaves like the gears of a clock was discussed by French philosopher and mathematician Rene Descartes (1596–1650), who developed it as a philosophy of scientific thinking. Isaac Newton (1643–1727) developed this further in his mechanical physical laws.

We will discuss some of the utility of this near-decomposable nested node structure when we turn to CAS-inspired actions in the next subsection.

Adaptation

Another essential property of CAS is that it *adapts* as agents within the system react to decisions made by other agents and in response to changing environmental conditions (Murphy, 2014). One result can be *self-organization*, in which a system's parts respond collectively to challenges without central coordination, resulting in a dynamic (as opposed to static) equilibrium (Garfinkel, Shevtsov, and Guo, 2017). Military units are organized based on doctrine and experience to achieve relevant ends within a context. Their behaviors can be expected to change in response to changes in many areas, including strategic priorities, adversary tactics, technology and equipment, resourcing levels, past performance, and other drivers. Although some of these changes will be the result of top-down orders, others are self-initiated and still others are the result of bottom-up behaviors. For example, a captain may change tactics because a bridge has been destroyed; the mission may change because of orders from a major, who got them from higher command; or the unit may stand down temporarily because so many soldiers in the command have begun to ignore orders, take cover, or shake out of fear.

The decisions and changes that drive adaptive restructuring of CAS tend to come from all levels of the network—top to bottom. This kind of adaptation can be particularly effective in a military context when clear strategic guidance comes from the top and best operational practices coming from other levels can be quickly identified and disseminated.

One aspect of adaptation is that it is *temporal* or occurs over time. A CAS takes time for lessons to be learned, structure to be adapted, and adaptation to propagate. Understanding a system's speed of organization and adaptation can be critical to understanding how it will respond to change and stress.

Emergent Properties

A CAS tends to have properties at higher scales that cannot be described in terms of the properties of the constituent components. For example, one person cannot be *at war* (although the term is used figuratively); war is a collective activity requiring many people working in concert with allies on the same side and in response to their enemies on the other side. War, therefore, is an emergent phenomenon that comes into being when two militaries engage in violent conflict. It has meaning when looking at the overall pattern of behavior in a way that it does not when looking at a single individual.

Phase transitions, known colloquially as "tipping points," are a major emergent property of many CAS (Morris et al., 2019). Phase transitions consist of a qualitative change where new systemic conditions emerge, such as new rules of new characteristics. Different phases operate by different rules, such as behavior of molecules in liquid or gas form. They can be precipitated by an event, such as the self-immolation that precipitated the Arab Spring or the assassination of

Franz Ferdinand that precipitated World War I. When a phase transition occurs, rules, culture, and expectations are different for both physical and social systems. When someone says "everything changed after 9/11," this sentiment expresses a concept of a social phase transition.

The boiling of water is a canonical example of phase transitions. The H_2O molecules in liquid phase each have a level of kinetic energy, but because they are tightly packed together, this energy can only be manifested as vibrational and rotational motion. However, at a temperature of 212° F (corresponding to a certain average kinetic energy level at a certain atmospheric pressure), the molecules no longer constrain one another and can also produce translational motion. This produces the dramatic transition from liquid to gas, during which the substance takes on radically different properties. If we observed a single molecule at different energy levels, we would see no evidence of a transition as it gained energy—boiling is a collective activity that only has meaning in the context of a large collection of molecules.

In more general CAS terms, phase transitions are abrupt changes in the qualitative behavior of a system in which the constituent parts of the system remain essentially the same, and the external stimulus on the system changes only slightly but the behavior of the system changes dramatically. A command and control (C2) system becoming overwhelmed might be an example in a military context, as would be the transition from a hostile standoff to a pitched battle.

Domains of operation can refer to the location in which conflict occurs, such as land, sea, and air, but also includes such categories as space, cyber, and information. Other useful groupings include basic warfighting functions (C2, movement and maneuver, intelligence, fires, sustainment, and protection), the division between officer and enlisted, and the division of the Air Force into wings, groups, squadrons, and flights. Although these groupings may be somewhat arbitrary, they generally derive from the principle that people and equipment with tightly integrated functions should be managed together and that each commander should have oversight of a (humanly) manageable number of subelements. So although the precise structure of the Air Force may be a product of historical development, the functional demands under which the service operates and the capacities of the people who staff it were bound to produce an organization with the same general character.

Complex Adaptive Systems–Inspired Actions: Characteristics That Provide Means of Attacking the Adversary's Decision Steps

By viewing conflict as a CAS, the complex features of an environment can be harnessed to create more-challenging conditions for an adversary or can be used to directly attack an adversary. We describe these as *complexity imposition* and *complexity attack*. As defined in Volume I, to *impose complexity* is to take an action that increases an aspect of the complexity of the environment in a way that makes it more difficult for an adversary to make decisions or operate or helps shape conditions to favor Blue. Thus, to conduct a *complexity attack* is to take an action that exploits characteristics of CAS to have a deliberate negative effect on the adversary.

10

We have identified four general categories of actions that gain their utility directly from the CAS nature of the system (Figure 2.2).[3] These are degrade operational picture, impair Red response, span organizational boundaries, and leverage non-linearities.

Figure 2.2. Avenues for Exploiting Complexity

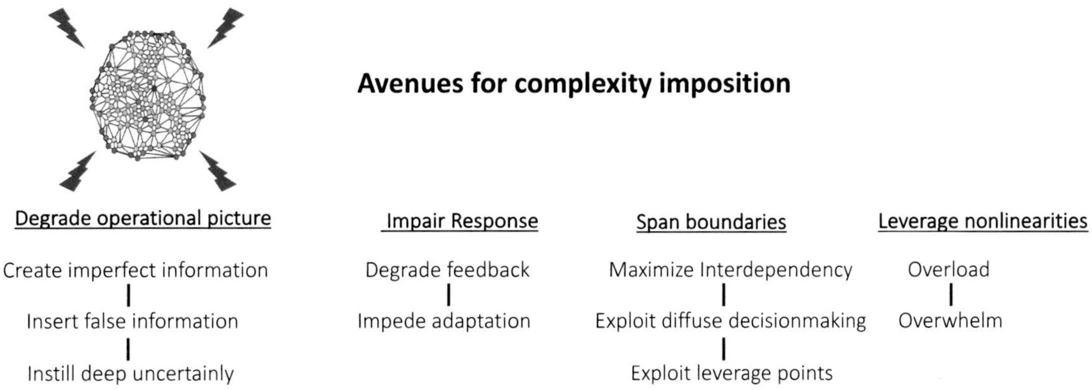

Avenues for complexity imposition

Degrade operational picture	Impair Response	Span boundaries	Leverage nonlinearities
Create imperfect information	Degrade feedback	Maximize Interdependency	Overload
Insert false information	Impede adaptation	Exploit diffuse decisionmaking	Overwhelm
Instill deep uncertainly		Exploit leverage points	

Degrade Operational Picture

Bounded rationality is a property of CAS that is especially relevant to human systems (Simon, 1972). This concept acknowledges that actors operating within a CAS environment cannot possibly know everything that is relevant to the decisions that they need to make nor could they make strictly optimizing sense of that information even if they had it. Because decisions are being made at all levels of an organization, and decisions depend on observed and anticipated adversary actions (which, in turn, depend on adversary observation and anticipation of blue actions), CAS decisionmakers are always operating with limited information and limited cognitive processing capacity. Although each side strives to know as much as possible and think as clearly as possible about its adversary, the CAS perspective acknowledges that such limitations are inherent and looks for ways to minimize the impact of these limitations on Blue and maximize impacts on Red.

Although the principle of bounded rationality implies that Blue forces can never make truly optimal decisions, it implies the same of Red forces. Assuming comparable Red and Blue forces—as long as Blue can maintain superiority in doctrine, C2, intelligence, and decision systems—the inherent bounds on rationality work to the advantage of Blue. We can establish a couple of categories for characterizing ways in which Blue can exploit bounded rationality.

First, Blue can maximize the asymmetry of known unknowns by creating *imperfect information* and ensuring that Red knows less than Blue does about what is happening. This

---

[3] These exclude actions, such as maintaining equipment with superior performance (in terms of range, power, reliability, and the like) compared with that of any adversary, that may be critically important but do not derive their motivation from the CAS perspective. This is essential but does not draw its desirability from a CAS perspective.

takes the form both of maximizing Blue access to information and of minimizing Red access to information. Blue access to information starts by developing strong systems for maintaining internal situational awareness. Blue decisionmakers should have access to very good information about Blue forces that is presented in the most digestible and actionable way as possible. This helps maximize the gap between what Blue knows about Blue forces and what Red knows about Blue forces. Blue can also expand this gap by maintaining superior intelligence about Red forces and capabilities.

As a second general class, Blue can exploit bounded rationality by feeding *false information* to Red—leading Red to make bad decisions or be slower in decisionmaking as it works to separate good from bad information. The history of armed conflict offers many examples of successful military deception, which can be productively analyzed through the CAS frame.

Although the first two classes of exploiting bounded rationality focused on denying information or feeding false information to the adversary, they take the form of maximizing the adversary's *known unknowns*: The adversaries know what they want to know but either cannot discover it or get it wrong. The creation of *deep uncertainty* is a third general class. Under deep uncertainty, an analyst, decisionmaker, or different parties to a decision are unable to agree on the nature of the problem or the objectives (Lempert, 2002; Lempert, Popper, and Bankes, 2003; Lempert et al., 2013). "Wicked problems" have this character (Churchman, 1967; Rittel, 1973). Blue can gain advantage by developing a clearer understanding of its own goals and objectives than Red and by developing a relatively better understanding of Red's goals and objectives than Red has of Blue's.

Blue can also exploit deep uncertainty by creating situations where Red does not understand how the system works. This may lead Red to focus its efforts (intelligence or otherwise) in areas that are not the most useful or to take actions that do not lead to the outcomes that Red actually desires. However, a caveat is needed because Blue does not always want Red to make bad decisions. In some cases, bad decisions by Red can lead to outcomes that are worse for both parties, such as when a country's weakening economy creates conditions for fostering terrorism.

Impairing Response

A CAS view provides perspective on how forces adapt to changing circumstances and changing adversaries. As just mentioned, the self-organized adaptation of human CAS involves observations and decisions made at many different levels of the organization. Some changes come from the top, others are pioneered in the field and then spread through the ranks, and others are a result of changing technology or other circumstances that may take time to be fully incorporated into doctrine.

Generally, it is advantageous for Blue forces to be quicker and more adaptable than Red. This can, however, prove challenging because the U.S. military is invariably larger and more mature than its adversaries, with more-developed processes and clearer procedures. Although these properties provide real advantages in familiar environments, they also create a degree of inertia, making it harder for the organization to change an adapt to new situations.

Feedback is a common feature of CAS and a key step in adaptation. *Gathering feedback* is the process of observing the outcomes and side effects (such as unintended consequences) of actions and using these outcomes to inform future actions (Brinsmead and Hooker, 2011; Grösser, 2017). This one of the primary functions of military intelligence, yet it is not always pursued as effectively as it might be. The 2006 *Iraq Study Group Report* observed the following of U.S. intelligence agencies:

> They are not doing enough to map the insurgency, dissect it, and understand it on the national and provincial level. The analytic community's knowledge of the organization, leadership, financing, and operations of militias, as well as their relationship to government security forces, also falls short of what policy makers need to know. (The Iraq Study Group, Baker, and Hamilton, 2006, p. 62)

This is an instance in which the U.S. military's substantial advantage in material and organization may have insulated it from the need to adapt, leading to insufficient gathering of feedback and problems in adapting to the emerging counterinsurgency environment in Iraq.

Even given proper feedback, adaptation remains challenging in a mature organization. Blue needs to establish systems for *making use of feedback* and adapting its structure, tactics, goals, and behavior to meet the challenges it faces (Diehl and Sterman, 1995; Rickles, Hawe, and Shiell, 2007). The CAS lens draws attention to the advantage that can be gained by doing this more quickly and effectively than Red forces can do the same thing.

Spanning Organization Boundaries

Conceiving of both Blue and Red as networks provides a means of analyzing avenues of attacking Red that Blue might use to gain advantage. The nested structure of these networks means that not all nodes have equal value. It also invites analysis of the interventions that might be most effective in preserving Blue capabilities and disrupting Red capabilities (Ruiz-Martin, Paredes, and Wainer, 2015).

Interdependencies create the opportunity for cascading impacts. A disruption in one node may create disruptions in a string of nodes that are dependent on that node. This, in combination with disruptions from other upstream nodes, may cause major problems for downstream nodes. This can both disrupt critical functions and lead to the shift of Red's internal environment to the point where Red needs time to assess and understand its own position, leading to slower and less effective response.

Diffuse decisionmaking is a characteristic of self-organization in human-based CAS. Systems tend to evolve to a point where decisions are made at all levels throughout the organization. This means that there can be multiple sources of authority and accountability with respect to some elements of the system and that no single person or group is functionally in charge of everything. On the one hand, this kind of decisionmaking structure can be extremely robust, allowing the organization to operate even when important nodes are disrupted. On the other hand, it can make decisions complicated when multiple decisionmakers who need to coordinate on a common action are involved.

A major implication of the hierarchal structure of military networks is that advantage can be gained by forcing the adversary into actions that require the coordination of nodes under a different command. A related action that is intended to be disruptive, requiring a response from a cohesive group that has trained together and answers to a single commander, is not as disruptive as an action that requires a response from several groups that are located in different places and different branches of the command hierarchy. Response in such a situation will require less efficient communication and decisionmaking that involves either the development of consensus or additional levels of command. In either case, the boundary-spanning response is likely to be slower and less effective.

Conversely, such actions on the part of Red should be anticipated, and boundary-spanning mechanisms should be put in place to handle these actions. These might take the form, for example, of new doctrine, the establishment of staff liaisons, and the conduct of joint training exercises. There is also opportunity to undermine Red's trust in its own networks and authorities, such as through information warfare campaigns.

CAS network structure can also be exploited by finding *leverage points* within it (Meadows, 1999; Russell and Smorodinskaya, 2018). The inherently uneven nature of both C2 networks and the network of unit interdependencies mean that there are bound to be some parts of the network where a relatively small disruption can have a large effect (Hofstetter, 2019).

In the case of C2, Red may have commanders or command systems that are critical to its decisionmaking. These make high-value targets for disruption. In identifying such leverage points, it is crucial to understand the nature of adversary C2. Formal chain of command is one thing, but peer-to-peer communications and information systems may provide critical links within adversary C2 that do not have the same structure as the chain of command. These links may provide resilience in the face of chain-of-command disruption but also may present vulnerabilities through the spread of confusing or inconsistent information.

Operational interdependency suggests a different network that may also present leverage points. Critical bottlenecks may be identified by tracing the parts of the adversary organization that depend on one another. As a simple example, Red may have a single fuel depot located at a port that serves land and air operations in a battle space. Even if the Army and Air Force use different fuels and manage them under different command, the physical colocation produces a common network node, the functionality of which is essential to the conduct of multiple operations across two services. This would be an extremely high-value node. Although this example is rather obvious (fuel depots are a classic point of attack), the network conception provides a general method to identify and rank high-value nodes.

The interaction of this operational dependency network with the command structure network presents opportunities to choose actions that require response from interdependent units that are closer in functional relationship than they are in command relationship. Response in such a situation will incur overhead as units work to coordinate their actions and develop consensus without direct command oversight. In cases where the need for such coordination is common or expected, boundary-spanning mechanisms can be expected to have evolved to facilitate such coordination. However, actions that require a response from groups that do not commonly work together may lead to delay, miscommunication, and other dysfunctional responses that work to the advantage of Blue.

Leveraging Nonlinearities

A side effect of the complex network structure, decentralized control, feedback, and adaptation of CAS is that they tend to contain areas of dynamic nonlinearity—parts of the behavior space where small changes in input can produce large (or even discontinuous) changes in output. These nonlinearities are one of the defining characteristics of CAS. The phase changes discussed earlier are examples of such nonlinearities (Richardson, Paxton, and Kuznetsov, 2017).

One example of a practical dynamic nonlinearity is the *overloading* of a node in the Red network. A commander might perform well under a light load and continue to perform well as the load increases, making only the occasional omission or error—with errors increasing in proportion to the workload. However, at some point, the information processing and communication capability of the commander becomes overloaded, and many more things start to get dropped. The falloff in performance beyond this point is likely to be quite steep because the commander is unable to handle or process additional information. In the longer term, the capabilities of the commander might be extended by creating a headquarters unit with well-defined ways of dividing the workload, but this takes time, and the transition cannot be made because the commander struggles to manage real-time events. The same logic applies to the overloading of any C2 system. Overloading, as described here, is an example of a phase change that is highlighted by the CAS lens.

A similar dynamic nonlinearity is displayed by the phenomenon of *overwhelming*. When Blue attacks and Red holds a material advantage, Red is likely to fight and win—possibly with light losses. When the two sides are evenly matched, the outcome is uncertain, but losses are likely to be heavy on both sides. When Red is at a slight disadvantage, Blue is more likely to win, with Red possibly taking heavy casualties. However, when Red is at a substantial disadvantage, it is likely to retreat (or surrender) immediately. In this situation, Red is overwhelmed by superior force and is not able to put up a viable defense. The difference between the case in which Blue has a small advantage and the case in which it has an overwhelming advantage may be minor, depending on how Red views its prospects. The difference in outcomes, however, is quite large.

Description and Application of a Complexity Lens

By conceiving of Air Force operations in terms of the properties just described, we are able to develop a CAS lens that enables an analyst or decisionmaker to identify whether, how, and to what degree a system or situation is complex. By delineating and measuring characteristics of CAS, it makes the concept more tractable for the Air Force or other armed forces.

Viewing operations or force planning through the CAS lens suggests various actions that take advantage of the CAS perspective. In Chapter 3, we present a tool to recognize the adversary's complexity challenges (if any) and to identify ways of exploiting them. The tool "sees" the adversary's organization, its adaptive mechanisms, and the emergent properties (if any). It then suggests vulnerabilities to look for and exploit.

3. Complexity Adaptive Systems Lens "User Manual"

Identifying the Complex Aspects of a System

This chapter supports the complexity lens rubric (Table 3.1) that has been developed to aid the AFRL in identifying the extent to which aspects of CAS are represented in programs, initiatives, and fielded systems.[4] In other words, the complexity lens rubric helps review S&T efforts to identify the extent to which aspects of CAS are represented (or not) in a research portfolio. We will review each line in the rubric, briefly discussing the relevant aspect of CAS and how it might be applied in the context of the complexity lens.

This rubric should be thought of in terms similar to the well-known doctrine, organization, training, materiel, leadership and education, personnel, and facilities (DOTMLPF) framework as defined in the U.S. Department of Defense's Joint Capabilities Integration Development System (JCIDS). Like the DOTMLPF, the rubric is designed to help analysts think of the various aspects of an enterprise and be sure that they are not missing important strategic considerations. Similarly, the CAS rubric is designed to help analysts see the extent to which a given system incorporates CAS principles and to understand what parts of the CAS space are more or less represented within a single system or across a set of systems.

The rubric is divided into three major sections based on fundamental properties of complex systems: *structural organization*, *adaptation*, and *emergent properties*. The properties are chosen to provide a minimal sufficient description of the properties of CAS in general. The rubric considers the extent to which the system has these properties, then asks about certain actions that the properties imply or enable. As reflected in Table 3.1 and in the discussion that follows, these actions are examined with respect to the impact they might have on the ability to defend ("defense") and the ability to attack ("offense").

---

[4] AFRL is the Air Force's S&T organization and the sponsor of the work reported here. AFRL Information Directorate was interested in a complexity rubric to evaluate the S&T research portfolio.

Table 3.1. Complex Adaptive Systems Lens Rubric

Structural Organization	Properties	System Has Characteristic?	
	Uses network structure?	3	
	Has multiple nested levels?	3	
	Avenues	**Defense**	**Offense**
	Maximize interdependencies	3	0
	Exploit diffuse decisionmaking	3	0
	Exploit leverage points	2	3
Information Flow	**Properties**	**System Has Characteristic?**	
	Depends on information flow?	3	
	Avenues	**Defense**	**Offense**
	Create incomplete information	3	1
	Insert false information	3	1
	Instill deep uncertainty	3	1
Adaption	**Properties**	**System Has Characteristic?**	
	Adaptive	3	
	Self-organized	3	
	Temporal	3	
	Avenues	**Defense**	**Offense**
	Gather feedback	3	2
	Make use of feedback	3	3
Emergent Properties	**Properties**	**System Has Characteristic?**	
	Abrupt changes	1	
	Domains of operation	1	
	Avenues	**Defense**	**Offense**
	Overload	3	2
	Overwhelm	1	2

Framing a System in Complex Adaptive Systems Terms

To understand the extent to which a given system is or is not complex, it is essential to frame it in appropriate terms. We found that it is often useful to think of the system as a type of network, potentially with many interacting parts and internal feedback loops. By definition, any complex system can be framed this way. In contrast, a simple or even complicated system will correspond to a simpler network with only limited internal feedbacks. For example, a diagram showing how a rifle works will be a set of linear steps, with few if any feedbacks or parallel processes. A diagram of conducting an air campaign, however, can easily be drawn as a complex web of people and equipment that are continually adjusting and adapting to one another. The rifle is a simple system, and the air campaign is a complex one.

The relevant network might be in many dimensions, including social, geographical, temporal, functional, and so forth. Before applying the rubric, it is critical that the analyst think through the various ways that the system under analysis might be framed in these network terms.

In the following sections, we unpack each of the properties and actions listed in the rubric, describing them abstractly and pointing out how they might be relevant to Air Force concerns. It may be helpful here to think through an example. Imagine a software tool under AFRL development designed to aid the planning of Blue air operations, in which thousands of daily sorties are scheduled rapidly. The tool would capture many of the planning factors that are currently manpower-intensive. For example, it would consider sortie packages, air refueling locations, air space deconfliction, take-off and landing schedules at various airfields, weather constraints, aircraft maintenance schedules, ingress and egress routes to target locations, weapon inventory and location, the joint integrated priority target list, and crew rest requirements. In this example, the rubric would show that the CAS system under consideration would be Blue planning. There would be little consideration to complexity attacks on Red, except possibly in the instance of electronic warfare support of mission packages. A future spiral version of the tool might add cyber and space planning to insert cyberattacks when an airborne mission package is most vulnerable or to plan imagery collections from satellites that are delivered to the cockpit for final positive identification of a mobile target. We reference the elements within the CAS rubric to see how the example software tool might be "rated" regarding complexity. The results are summarized in Table 3.1.

Within each major section, the rubric asks the analyst to think about the actions that the system defends against or enables. This is intended to focus on the extent to which the system is leveraging CAS principles in the actions it is designed to carry out. There are many ways for a system to be successful—this rubric is designed to highlight the extent to which CAS principles are central to the design of the system, not to evaluate its overall worth.

Property: Structural Organization

CAS are generally organized as networks, with material, information, and/or decisions moving between interconnected parts. These parts are often connected in multiple ways, enabling feedback and alternative pathways through the system. Again, this network may take on many forms, ranging from a human C2 system to a computer data network to a manufacturing supply chain. The following indicators probe different aspects of how the system is organized.

To what extent does the system involve a *network structure*? Is it a simple network or a complicated one (0 = network not relevant; 1 = basic network, not fundamental to system; 2 = network is essential part of system; 3 = complicated, system critical network)?

The networks that compose a complex system are often organized in a multilevel, nested structure. These may include modular components that have complex internal connections and simpler connections to other modular components (such as the internal organ structure of living organisms, a decentralized computing network, or an interacting set of specialized military units).

To what extent does the system involve *multilevel, nested network structures* (0 = simple network topology; 1 = network with some variety in node functions; 2 = nested, multilevel network; 3 = dynamic, nested, multilevel network structures)?

Returning to our Blue airborne planning tool, we see that the tool incorporates information across logistics, weather, multiple aircraft types (supporting and supported), and maintainers and pilots—in other words, the network is complicated. The network structure would be rated 3. There may be multiple levels represented in the tool, where mission packages and supporting resources are formed by multiple squadrons and multiple locations under different commands; this all may be changing over time. Blue is organized with dynamic, multiple nested levels, so it would be rated 3.

Avenues for Complexity Imposition Related to Structural Organization

A network structure can have strengths and weaknesses. A well-designed network can be both efficient and resilient. Disruption of some nodes, however, may have impact or influence across different parts of the system. Some actions may have impacts on nodes that are not closely connected. This section explores the ways in which the system uses network structure properties. We think not simply of computer or communication networks but all of the various ways information or material might move within the system.

Maximize Interdependencies

The parts of a CAS are interdependent, and these interdependencies operate across a range of scales, distances, and levels of directness. These interdependent structures allow for specialization and flexibility; however, they can also produce weaknesses when a response depends on a chain of actions that is not sufficiently robust or resilient.

- Defense: Does the system have interdependent components (if not, 0 as not applicable)? If so, are these interdependent parts designed to work together in a robust, resilient way (1 if marginally so, 2 if this is a design feature, 3 if dynamic resilience is integral)?
- Offense: Does the system require the adversary to respond with multiple, interdependent systems (0 if a single response is adequate; 1 if multiple, but uncoordinated, responses are needed; 2 if multiple coordinated responses are needed; 3 if coordinated responses are required from adversary components that do not customarily coordinate)?

The Blue planning tool would have interdependent components, including tankers to fuel short-range fighter aircraft traveling from long distances or aircraft maintenance information for sortie package availability. The tool would rate 3 for the defense question above but 0 for the offense question.

Exploit Diffuse Decisionmaking

Many CAS involve decentralized decisionmaking. This can provide robustness in the face of losses and disruption in communications but can also lead to inefficiency if these decisions cannot be well coordinated.

- Defense: Does the system incorporate diffuse or decentralized decisionmaking as part of its design?
- Offense: Is this system designed to take advantage of diffuse adversary decisionmaking by, for example, disrupting adversary ability to coordinate decisions?

The rating of our Blue tool example would depend on the degree to which the tool accounts for decentralized or distributed planning. If concepts of adaptive basing and smaller multidomain operations centers are accommodated in the tool, the rating would be perhaps 3. Here, expert judgment must be used to rate the degree to which diffuse decisionmaking is considered for Blue. The tool does not consider the adversary decisionmaking, so the rating would be 0 for that.

Exploit Leverage Points

Some network structures include nodes that are particularly critical to their operation. This might be a fuel depot that supplies regional field operations, a central command headquarters, or a computer communications hub that is not part of a redundant architecture.

- Defense: Does the system have nodes in its network that are particularly vulnerable to disruption?
- Offense: Does the system attack nodes in the adversary network that are particularly vulnerable to disruption?

The planning tool example is helpful in working to identify the critical paths for airborne mission planning and to seek mitigation measures to address those vulnerabilities. For example, can data be mirrored locally to account for periodic outages in communications? Can the tool use a predictive algorithm to reduce uncertainty when the link is down? Are the data verified in some way to ensure against corruption? Thinking about the ways that the Blue tool can decrease vulnerabilities may also help researchers come up with new ways to attack the adversary's system that performs similar tasks. If the tool is being built with consideration of these factors but relies on other systems to protect vulnerable features, it might rate 2 for defense. If the system includes tools for prioritizing targets based on their importance to overall adversary operations, it might rate 3 on offense.

Property: Information Flow

In most CAS, the component parts of the system (whether they be people or pieces of equipment) have neither perfect information about their adversary and environment nor perfect reasoning ability to process the information that they do have given the constraints of time, memory, and the like. Operational efficiency requires that good information be collected and processed by the system. These questions explore complexity impositions on the information environment.

To what extent does this system depend on the collection, communication, and processing of information (0 = low need for information, 1 = local data collection, 2 = data collection and sharing between different parts of the system, 3 = complex data collection, processing, and communication are essential to system operation)?

The information flow is clearly related to the vulnerability potential just discussed for the Blue example. With distributed bases and decentralized execution of airborne operations, the system is heavily reliant on the collection, communication, and processing of information and would likely be rated 3.

Avenues for Complexity Imposition Related to Information Flow

Create Imperfect Information

To what extent does the system operate to gain advantage from incomplete information?

- Defense: Is the system designed to protect Blue information? Is the denial of information to the adversary a major design goal (e.g., stealth, encryption, covert activity)?
- Offense: Is the system designed to gather or obtain information? Is discovering information about the adversary a major design goal?

For our example, the information-protection measures taken are a critical design consideration for the Blue planning tool from the start. From the beginning, has the tool worked on security compliance? If so, this might be rated 3 for defense. If intelligence operations are not part of the system design, but the tool does involve keeping track of what is known about the adversary air defenses, it might be rated 1 for offense.

Insert False Information

To what extent does the system deal with false information?

- Defense: Is the system designed to reject false information provided by the adversary?
- Offense: Is the system designed to provide false information to the adversary?

The veracity of data in the Blue example tool is critical to effective planning. What are the checks on the data to ensure that they are not tampered with? If strong design principles are being used to ensure that data are not tampered with, defense might be rated 3. If the system is designed to support deceptive air operations, but does not directly generate false information, it might be rated 1 for offense.

Deep uncertainty in complex systems means that the actor not only does not know important things but does not even know what needs to be known. The actor does not know the rules of the system and is not able to accurately anticipate the outcome of its own actions. Examples might include strategic shifts, randomized actions, or other measures designed to create an uncertain decisionmaking environment.

- Defense: Is the system specifically designed to maintain clarity and reduce the probability of deep uncertainty?
- Offense: Is the system specifically designed to keep the adversary uncertain about the workings, capabilities, and intentions of Blue?

If the primary purpose of the system is to maintain situational awareness and coordination of Blue activities, it might rate 3 on defense. If it is not specifically designed to impose unpredictable actions on the adversary, it might rate 1 on offense.

Property: Adaptation

Adaptation is an essential feature of CAS. This section examines how the system incorporates adaptation in the face of changing circumstances.

To what extent is adaptation a part of the system? Is there reaction to the environment, or does the system always do the same thing (0 = nonadaptive, 3 = highly adaptive)?

To what extent does any adaptation have a bottom-up character? Are changes determined centrally, or is change decentralized, with subcomponents reacting and adapting independently to make the overall system a better fit for its environment (0 = change is centrally directed, 3 = subcomponents are highly autonomous)?

To what extent is the system designed to adapt more quickly than the adversary system (0 = speed is not a major issue for this system, 3 = adaptation speed is a critical design principle for the system)?

If the example tool took inflight changes—such as a tanker dropping out because of a mechanical problem, and then being dynamically replaced with another that happens to be nearby and tasked originally to a lower-priority mission—then its adaptation rating might be 2. If the tanker's status is part of an overall replanning cycle, involving the reassessment of all flights, then it would be top-down organization, rating 0. If the tanker's status is determined dynamically using only the affected airframes and systems, then it would rate 3 in terms of bottom-up organization.

The goal of a Blue planning tool like this one is to adapt as quickly as possible to the changing environment given the constraints of physical systems (e.g., time of flight versus distance, maintenance times). The tool should rate 3 if adaptation speed is a critical design principle for the system.

Avenues for Complexity Imposition Related to Adaptation

Gather Feedback

To adapt to circumstances, the system needs to gather feedback on what is happening and what is resulting from its current actions.

- Defense: To what extent does the system monitor its own functioning?
- Offense: To what extent does the system gather information about the adversary to evaluate the impact of its previous interactions?

The ability of the example Blue tool to self-monitor and assess functionality is something that must be planned for in the designing of the tool. This tool is primarily dedicated to monitoring Blue operations and detecting problems with them. It would thus rate 3 in defense. The tool might also serve as a means of distributing, but not gathering, information on the success of Blue operations and the current state of Red capabilities. This might be reflected with a rating of 2 for offense.

Make Use of Feedback

Having gathered feedback, the system must make changes and adjustments to put that feedback into action in adapted behavior.

- Defense: How effectively is the system able to use information about changing circumstances to protect itself and maintain function?
- Offense: How effectively is the system able use information about changing circumstances to shift its behavior to inflict harm on the adversary?

As with the capability to gather feedback, the ability to do something useful in a timely manner with that feedback is an essential consideration in the design of the example tool. If this is lacking, the researcher should consider adjusting the design or plan for a spiral change to incorporate it. This tool should be designed to rapidly replan missions in the face of shifting capabilities and circumstances. This might include something like an aircraft becoming unexpectedly unavailable. Building around this capability would rate 3 on defense. Assuming the tool is also designed to rapidly reprioritize targets as they change, the tool would also rate 3 on offense.

Emergent Properties

CAS tend to have properties at higher scales that cannot be predicted or understood in terms of the properties of the constituent components. This section explores some emergent features of the system and invites analysis of additional emergent features.

Abrupt Changes (Phase Transitions)

Some systems are designed to deliver consistent performance over a broad range of operating conditions. Others, however, are designed to shift suddenly from one mode to another.

- Does the system have distinct operational phases (0 = system behaves consistently; 1 = system has multiple modes, shifting under operator command; 2 = system can autonomously shift between multiple modes; 3 = system autonomously and dynamically changes behavior abruptly)?

In our planning tool example, there may be times when the tool is used for maintenance and training, other times when it is used to direct active combat operations, and still other times when it shifts from a primarily offensive to a primarily defensive mode. The tool would tend to behave consistently across these multiple modes, so rates 1.

Domains of Operation

Domains of operation do not emerge during the operation of a particular system but represent categories of equipment and responsibilities that emerge from operations over time. Domains of operation emerge over the course of decades and become part of the structure of military operations. This contrasts with such phenomena as battles that emerge over short timescales (e.g., minutes).

Land, sea, and air are the most basic domains, but at the same level, we also have such categories as space, cyber, information, and the like. At lower levels, other domains might be identified with the Air Force core missions of air and space superiority; intelligence, surveillance, and reconnaissance (ISR); rapid global mobility; global strike; and C2. For example, operations in one domain (e.g., space) may necessitate operating in another domain (e.g., cyber).

Does the system operate across multiple domains (0 = system is largely confined to a single domain; 1 = system operates across more than one domain; 2 = system integrates domains that are not commonly integrated; 3 = system is unusual in its ability to operate across domains)?

Our example Blue planning tool might rate low for operating across multiple domains because it primarily focuses on air operations. In this instance, it would rate 0. However, a case could be made that if it considered electronic warfare measures, such as jamming in mission package planning, then it may be rated 1.

Avenues for Complexity Imposition Related to Emergent Properties

Leveraging Nonlinearities

A side effect of the complex network structure, decentralized control, feedback, and adaptation of CAS is that they tend to contain areas of nonlinearity, in which small changes in input can produce large (or even discontinuous) changes in output. These constitute another broad class of emergent system behavior. *Overloading* and *overwhelming* are two examples of such behaviors.

Overload

In many cases, the performance of a system falls off quite abruptly as it reaches some sort of saturation point. An aircraft controller might be able to handle a large workload with few errors, but there comes a point where the controller is overloaded, and the rate of errors increases dramatically.

- Defense: To what extent is the system robust to overload (0 if this does not apply)?
- Offense: To what extent is the system designed to overload the adversary?

Whether the example Blue tool is robust to overload is something that can be tested during development, but prior to that, it should also be considered in the design phase. If the tool is designed to be scalable and is able to handle massive operations, it might be rated 3 on defense. If it is designed to facilitate operations of sufficient scale to potentially overload the adversary defenses, it might rate 2 on offense.

Overwhelm

Overwhelming takes place when one party perceives its disadvantage to its adversary to be so large that resistance is futile, pursuing instead such strategies as retreat, surrender, or simply absorbing the blow.

- Defense: Is the system robust to being suddenly overwhelmed? Does it behave per expectations when it is outmatched (0 if this does not apply)?
- Offense: Is the system designed to present the adversary with overwhelming force (which need not be kinetic)? Does the system seek to make use of the concept of overwhelming the adversary?

The ability to overwhelm the adversary based on the Blue tool is likely rated similarly to overload. The system may be vulnerable to a situation in which an adversary is able to present more threats that occur simultaneously than its designers ever anticipated. It would rate 1 on defense because it is strong under most circumstances but is foreseeably vulnerable here. A positive outcome for Blue, with thousands of daily military sorties, is to overwhelm the adversary, thereby helping it conclude that it is futile to fight. However, the planning system is not completely responsible for this overwhelming force. Therefore, the offense rating may be 1 or 2.

Summary

This chapter described a complexity lens rubric for reviewing S&T efforts to identify the extent to which aspects of CAS are represented (or not) in a research portfolio. The lens was exercised by using a hypothetical software tool for planning of Blue air operations in which thousands of daily sorties are scheduled. In this example, the rubric would highlight many aspects of CAS being applied to the process of Blue planning while also showing that the system design gives little consideration to complexity attacks on Red.

The hypothetical software tool aims to consider CAS characteristics of Blue's C2 as a network and as containing multiple levels. The tool should also consider adaptation characteristics—the ability to change in response to unfolding conditions, self-organization (e.g., use of mission type orders), and the timeliness of changes (or responsiveness). The ability of the planning tool to adjust to campaign phase transitions and incorporate operations from other domains would further leverage CAS characteristics. The tool must also consider complexity attacks by Red—for example, planning under uncertainty when Red has degraded the operational picture. Other areas to address via building mitigation measures in the tool include Red attacks on the ability to leverage feedback or to attack planning seams or boundaries between different planning groups. Red may also overload the tool with data, so mechanisms to address these possible vulnerabilities should also be a part of the research plan.

This is just one example of a research project that can be evaluated with a complexity lens rubric. The rubric can also be applied across a larger research portfolio. An analyst could sum scores across a range of projects to help identify where the overall emphasis resides and provide S&T decisionmakers with a framework for reviewing and adjusting emphasis based on their goals and objectives.

4. Markov Chain Formulation and Analysis of Decision Processes

In this chapter, we outline the mathematical material that supports the adversary decision flow in Chapter 3 of Volume I. We use properties of graphs and Markov chains to study the structure of the decision processes that are presented in Volume I. A *Markov chain* is a set of transitions, which are determined by some probability distribution, where the chain's process is characterized by its memorylessness (Soni, 2018). In this application, the adversary decision process is essentially a set of decision states and transitions that can be represented as a Markov chain. By viewing a Markov chain as a stochastic process on a graph, we can study the structure of the aforementioned decision processes at two levels of resolution. We will show that, with sufficient data, the Markov chain framework can be used to calculate the outcome of the respective decision processes. However, if data are insufficient to perform these calculations (which is often the case), we can still use the underlying graph structure of the Markov chain to comment on the structure of the decision space.

This chapter is structured as follows: First, we provide preliminary definitions that allow us to develop the necessary framework to discuss graphs and Markov chains. Then we show how transition matrices are constructed from Markov chains and present a few important properties of these matrices. We then present additional properties of Markov chains and graphs that are useful for analyzing properties of decision processes and outline a modeling approach.

Definitions

Throughout this chapter, we will use two example processes, referred to as Graph A and Graph B. These are shown in Figure 4.1 and Figure 4.2, respectively.

Figure 4.1. Graph A

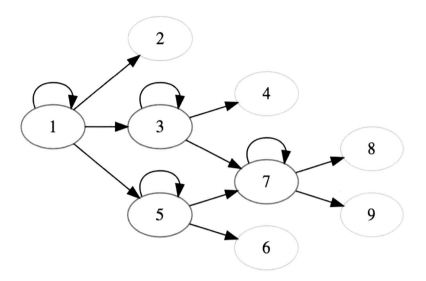

Graph A shows nine vertices and is associated with the straightforward strategic first-strike decision process described in Chapter 3 of Volume I. It includes Blue adding resiliency measures, such as active defenses or runway repair kits.

Figure 4.2. Graph B

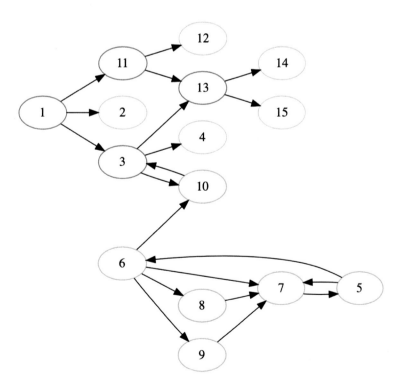

Graph B shows 15 vertices and is associated with the complexity attack on the adversary's first-strike decision process in Chapter 3 of Volume I, in which decoys and counter-ISR measures are employed by Blue forces at agile combat employment locations.

We will now give a few definitions related to graphs to fix terminology and notation.

- Let $V := \{v_1, \dots, v_n\}$ be a finite set of elements, which will be referred to as *vertices* (or nodes), and let $V \otimes V$ denote the set of ordered pairs (v_i, v_j) of elements in V.

  - The set of vertices for Graph A is $V_A = \{v_1, v_2, v_3, v_4\}$.
  - The set of vertices for Graph B is $V_B = \{v_1, \dots, v_{15}\}$.

- A relation on the set V is any subset $E \subset V \otimes V$. The relation E is symmetric if $(v_i, v_j) \in E$ implies $(v_j, v_i) \in E$ for all $i, j \in \{1, \dots n\}$. The relation E is said to be antireflexive if $(v_i, v_j) \in E$ implies $v_i \neq v_j$ for all $i, j \in \{1, \dots, n\}$.

  - A relation on V_A that describes the edges of Graph A is given by $E_A = \{(v_1, v_2), (v_1, v_3), (v_2, v_4), (v_3, v_4), (v_4, v_5)\}$.
  - A relation on V_B that describes the edges of Graph B is given by $E_B = \{(v_1, v_2), (v_1, v_3), (v_3, v_4), (v_3, v_5), (v_5, v_6), (v_6, v_7), (v_6, v_8), (v_6, v_9), (v_6, v_{10}), (v_5, v_7), (v_7, v_5), (v_8, v_7), (v_9, v_7), (v_{10}, v_{13}), (v_1, v_{11}), (v_{11}, v_{12}), (v_{11}, v_{13}), (v_{13}, v_{14}), (v_{13}, v_{15}), \}$.

- An undirected graph $G = (V, E)$ is an ordered pair where V is a set of vertices, and E is a set of symmetric antireflexive relations on V (in particular, $E \subset \{\{v_1, v_2\} \mid v_1, v_2 \in V, v_1 \neq v_2\}$). An element of the set E is referred to as an edge.

  - Let E_A' be the edge set that is generated by including all symmetric relations missing from the set E_A. Then $A' = (V_A, E_A')$ is an undirected graph.
  - Let E_B' be the edge set that is generated by including all symmetric relations missing from the set E_B. Then $B' = (V_B, E_B')$ is an undirected graph.

- An edge $e = \{e_1, e_2\} \in E$ is said to be adjacent to the vertex $v \in V$ if $e_1 = v$ or $e_2 = v$.
- Two edges $e = \{e_1, e_2\}, f = \{f_1, f_2\} \in E$ are said to be adjacent if $e_i = f_j$ for any $i, j \in \{0, 1\}$.
- A directed graph (or digraph) $G = (V, E)$ is an ordered pair where V is a set of vertices, and E (the edge set) is a set of antireflexive relations on V, which is not necessarily symmetric.

  - The Graph $A = (V_A, E_A)$ is a directed graph with vertices V_A and edge set E_A.
  - The Graph $B = (V_B, E_B)$ is a directed graph with vertices V_B and edge set E_B.

A Markov chain is a sequence of random variables $\{X_t\}_{t \in T}$, where T is finite or countable index set, such that

$$P(X_t = x \mid X_1 = x_1, \dots, X_{t-1} = x_{t-1}) = P(X_t = x \mid X_{t-1} = x_{t-1}),$$

given that $P(X_1 = x_1, \dots, X_{t-1} = x_{t-1}) > 0$. That is, the probability of the chain arriving in the current state only depends on the previous state.

We will also assume that the Markov chains we work with are time-homogeneous; that is, each of the probabilities in the memoryless property are independent of time. In the case of a Markov chain, the set S comprises the possible values of the random variables $X_1, X_2, \dots$, and is referred to as the *state space*. A directed graph can present the state space of a Markov chain.

Transition Matrix of a Markov Chain

Let $X = (X_t)_{t \in T}$ be a Markov chain with the state space realized as a directed graph $G = (V, E)$. Assume the graph G has n vertices; then the transition matrix for this Markov chain is given by $P = [p_{ij}]_{i,j \le n}$, where

$$p_{ij} = \begin{cases} P(X_t = X_j | X_{t-1} = X_i), & (v_i, v_j) \in E \\ 0, & \text{otherwise.} \end{cases}$$

Note that, for all pairs of vertices not connected in the underlying graph G, the corresponding entry in the transition matrix is zero because

$$P(X_t = X_j | X_{t-1} = X_i) = 0 \text{ if } (v_i, v_j) \notin E.$$

As an example, the transition matrix for a Markov chain on Graph A can be written as:

$$P_A = \begin{bmatrix} 1 - p_{12} - p_{13} - p_{15} & p_{12} & p_{13} & 0 & p_{15} & 0 & 0 & 0 & 0 \\ 0 & 1 & 0 & 0 & 0 & 0 & 0 & 0 & 0 \\ 0 & 0 & 1 - p_{34} - p_{37} & p_{34} & 0 & 0 & p_{37} & 0 & 0 \\ 0 & 0 & 0 & 1 & 0 & 0 & 0 & 0 & 0 \\ 0 & 0 & 0 & 0 & 1 - p_{56} - p_{57} & p_{56} & p_{57} & 0 & 0 \\ 0 & 0 & 0 & 0 & 0 & 1 & 0 & 0 & 0 \\ 0 & 0 & 0 & 0 & 0 & 0 & 1 - p_{78} - p_{79} & p_{78} & p_{79} \\ 0 & 0 & 0 & 0 & 0 & 0 & 0 & 1 & 0 \\ 0 & 0 & 0 & 0 & 0 & 0 & 0 & 0 & 1 \end{bmatrix}.$$

The entries in the matrix P_A are color coded to match the vertices of Graph A, as shown in Figure 4.1. In particular, the red entries correspond to nonabsorbing states, and the gray entries correspond to absorbing states.

The transition matrix for a Markov chain on Graph B, denoted by $\boldsymbol{P}_B$, can be written as:

$$
\begin{pmatrix}
-p_{12}-p_{23}-p_{111} & p_{12} & p_{13} & 0 & 0 & 0 & 0 & 0 & 0 & 0 & p_{111} \\
0 & 1 & 0 & 0 & 0 & 0 & 0 & 0 & 0 & 0 & 0 \\
0 & 0 & 1-p_{34}-p_{310}-p_{313} & p_{34} & 0 & 0 & 0 & 0 & 0 & p_{310} & 0 \\
0 & 0 & 0 & 1 & 0 & 0 & 0 & 0 & 0 & 0 & 0 \\
0 & 0 & 0 & 0 & 1-p_{56}-p_{57} & p_{56} & p_{57} & 0 & 0 & 0 & 0 \\
0 & 0 & 0 & 0 & 0 & 1-p_{67}-p_{68}-p_{69}-p_{610} & p_{67} & p_{68} & p_{69} & p_{610} & 0 \\
0 & 0 & 0 & 0 & p_{75} & 0 & 1-p_{75} & 0 & 0 & 0 & 0 \\
0 & 0 & 0 & 0 & 0 & 0 & p_{87} & 1-p_{87} & 0 & 0 & 0 \\
0 & 0 & 0 & 0 & 0 & 0 & p_{97} & 0 & 1-p_{97} & 0 & 0 \\
0 & 0 & p_{103} & 0 & 0 & 0 & 0 & 0 & 0 & 1-p_{1013} & 0 \\
0 & 0 & 0 & 0 & 0 & 0 & 0 & 0 & 0 & 0 & 1-p_{1112}-p_{1113} \\
0 & 0 & 0 & 0 & 0 & 0 & 0 & 0 & 0 & 0 & 0 \\
0 & 0 & 0 & 0 & 0 & 0 & 0 & 0 & 0 & 0 & 0 \\
0 & 0 & 0 & 0 & 0 & 0 & 0 & 0 & 0 & 0 & 0
\end{pmatrix}
$$

The entries in the matrix P_B are color coded to match the vertices of Graph B, as shown in Figure 4.2. In particular, the red entries correspond to nonabsorbing states, the gray entries correspond to absorbing states, and the green entries correspond to states that occur during complexity attacks.

We note that the transition matrix P is right-stochastic. That is, all elements are nonnegative, and the entries in each row sum to one:

$$\sum_{j=1}^{n} p_{ij} = 1.$$

The transition matrix can be written in the form $P = A D A^{-1}$, where D is a diagonal matrix with the eigenvalues of P along the diagonal. Assuming the chain starts in position $P_1 = P$, the distribution of the chain at time $t \in T$ can be computed as

$$P_t = AD^t A^{-1}.$$

We will now explain how the transition matrix can be used to find the steady state(s) of a Markov chain. This is a useful process for identifying the terminal nodes in the decision processes that we study in this report.

A vector $\pi = \left[\pi_j\right]_{1 \leq j \leq n}$ is called a stationary vector if $\pi_j \in [0,1]$ for each $j \in \{1, \dots, n\}$, and $\sum_{j=1}^{n} \pi_j = 1$, and

$$\pi_j = \sum_{i=1}^{n} \pi_i \, p_{ij},$$

for all $j \in \{1, \dots, n\}$. The final expression above can be written in matrix form as

$$\pi P = \pi,$$

where P is the transition matrix for the Markov chain. From this expression, we observe that the stationary distribution(s) for a Markov chain correspond to the normalized left-eigenvectors of the transition matrix P. The stationary distribution(s) of the matrix P can be interpreted as the long-time steady state(s) of the system. If all of the probabilities in the transition matrix are known, then this process tells us which states the Markov chain will converge to if there are absorbing states (equivalent to terminal nodes in the decision process).

Additional Properties of Graphs

Often all of the transition probabilities will not be known for the decision processes we are studying in this report. In this case, there are still a number of properties that can be deduced from the structure of the state space of the Markov chain by analyzing the underlying structure of the directed graph on which the process is defined. In this section, we will mention some of these properties.

The adjacency matrix of a graph $G = (V, E)$ shows how each node is connected and is defined as the matrix $A = [a_{ij}]_{i,j \leq n}$, where the components are given by

$$a_{ij} = \begin{cases} 1, & (v_i, v_j) \in E \\ 0, & (v_i, v_j) \notin E. \end{cases}$$

There are several graph statistics that can be used to quickly quantify the size and topology of a graph, and hence give some basic information about a graph's complexity. Many of these can be efficiently computed using the adjacency matrix. A sample of these statistics are given as follows:

- The *average node degree* is the average of the number of connections (degrees) of all nodes in a graph. This measures the average connectedness of a graph.
- The *degree centrality* is the fraction of nodes a given node is connected to in a graph. This measures how connected each node in a graph is with respect to the rest of the graph.
- *Eigenvector centrality* is the centrality of a node with respect to the centrality of its neighbors, using the eigenvector decomposition of the graph adjacency matrix.
- *Betweenness centrality* is the fraction of all shortest paths, for all node pairs, that pass through a given node.
- The *clustering coefficient* is the fraction of all possible triangles that exist with each node's neighbors.
- The *diameter* of a graph is the length of the longest of the shortest paths between any two vertices in a graph. More precisely, let (u, v) be a pair of vertices, and let $d(u, v)$ be the graph distance, which is the shortest path between the vertices u and v. The diameter of a graph G is then given by

$$Diam(G) = max_{u,v \in V} d(u, v).$$

The diameter of a graph gives an intuitive way of quantifying how large a graph is. Generally, graphs that are larger and have more connections tend to be more complex.

- The *average distance* for a graph is the mean of all pairwise distances $d(v_i, v_j)$ between vertices v_i and v_j for all $i, j \in \{1, 2, \dots, n\}$ for a graph with n vertices. The average distance can be computed by taking the average of the values in the graph distance matrix defined by $M_{Dist} = [d(v_i, v_j)]$. Although the diameter gives extremal information about the largest separation between points, the average distance gives information about how separated vertices are on average. Generally, the larger the average distance is, the more complex the graph will tend to be.

- A cycle is a series of vertices, starting and ending at the same vertex, such that each two consecutive vertices are adjacent to each other in the graph. The *number of cycles* in a graph can help to distinguish complex decision processes because Markov chains on directed graphs with cycles are often accompanied by feedback, a characteristic of complexity.

- The *girth* of a graph is the size of the largest cycle in the graph. The girth of a graph gives information about how a graph is connected and is a dual idea to the connectivity of a graph. In particular, the girth of a graph is the k-connectivity number for the dual graph. In the context of decision processes, cycles add additional processing time and therefore slow down the decisionmaker. Therefore, smaller cycles (if any) are more optimal.

- *Terminal nodes* in a graph can be identified with the adjacency matrix by identifying rows in which the only entry is in the diagonal. These are particularly important for analyzing decision flows because the flow will terminate at these nodes in the graph.

- The *number of paths to terminal nodes* can be used to estimate the relative likelihood of arriving at each terminal node (assuming uniform probabilities for an overlayed Markov chain).

Each of these statistics gives information about different aspects of a graph's topology, which is relevant to decision processes in the warfighting domain.

Modeling Approach

Our analytical approach to the decision flow diagrams laid out in Chapter 3 of Volume I models this decision process by conceiving of it as a Markov chain decision tree, in which there is a probability associated with the transition from each decision state to the next. In the context of our first-strike example, each decision probability is the likelihood that a decisionmaker will view the outcome of a given step as favorable and move on to the next step in the decision flow. To formalize our analysis, we can present the decision framework as a transition matrix, as defined previously in this chapter. To make this more concrete, we will synthesize previous calculations from this chapter to show how this analysis is performed in the case of the adversary first strike (or agile combat employment concept of operations) that is described in Chapter 3 of Volume I. The decision flow for this example is shown in Figure 4.3.

Figure 4.3. Decision Flow for Strategic First Strike (or Agile Combat Employment Vignette) Described in Chapter 3, Volume I

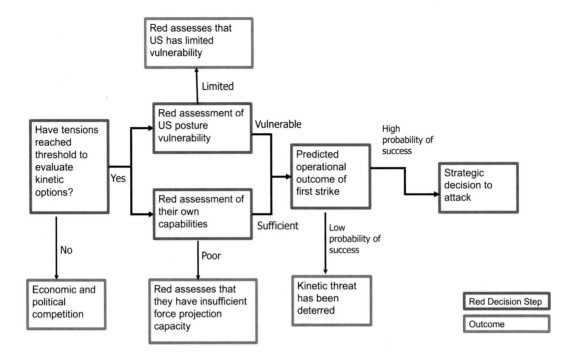

Graph A from Figure 4.1 is a representation of the decision process in Figure 4.3. The corresponding transition matrix is given by $\boldsymbol{P}_A$, which was derived previously in this chapter. By defining this decision framework as a Markov process, we can identify the ways that complexity attacks reshape adversary behavior and analyze relative impacts of different potential deterrence strategies.

In parallel to this decision flow, the impact of base survivability efforts can also be viewed as a transition matrix, in which these resiliency measures reduce the probability associated with adversary's assessment of U.S. posture vulnerability (Figure 4.4).

Figure 4.4. Transition Matrix Associated with Increased Base Resiliency

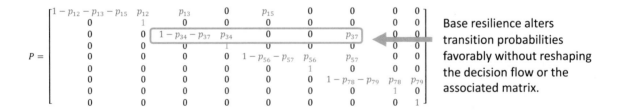

In contrast to this case, the use of decoys and counter-ISR measures described in Chapter 3 of Volume I reshapes the decision flow by adding fundamentally new elements to it. This reshapes both the graph and the associated transition matrix (Figure 4.1 and Figure 4.3). In this case, an adversary's likelihood of attacking is reduced by denying it the ability to assess the potential impact of a first strike. Graph B is the corresponding digraph, and P_B is the resulting transition matrix. The transition matrix associated with this decision flow reflects the added complexity of the decision calculus.

By taking the limit of the matrix given infinite time, it is also possible to compute the stationary state of the chain, which gives us the probability of the chain converging to each of the potential end states of the decision tree. These outcomes can then be directly compared with the impacts achieved by manipulating specific transition probabilities to weigh the operational impacts of different concepts of operations.

Topological Interpretations of Complexity Attacks

Although calculating the outcomes associated with a transition matrix is a powerful approach, it may not be practical in many cases because the values associated with specific transition probabilities may not be known. Although transition probabilities can be estimated through a variety of approaches ranging from expert assessments and intelligence to wargaming and simulation-based approaches, this may not be practical at large scale. It is difficult, for example, to assess the likelihood of a deception campaign's success or the probability that an adversary decisionmaker will prefer to delay a decision to re-collect imagery so that the outcome of a matrix can be computed correctly. In the absence of this information, we have identified several additional approaches that allow the impacts of complexity imposition to be described even in the absence of measured transition probabilities.

Structurally, the degree of decision branching induced by a complexity attack can be measured by observing the number of off-diagonal elements introduced to the transition matrix. Simple decision processes often have a linear Markov chain structure, which corresponds to a bidiagonal transition matrix. More-complex decision processes have more off-diagonal components, which imply feedback and/or a bifurcation in the decision space with multiple possible outcomes. These structural properties are likely to slow decision processes because there are more assessments that must be completed to arrive at a given end state.

The ways that complexity reshapes an adversary's decisions can also be modeled by overlaying a Markov chain structure onto the digraph to map the flow as a set of decision processes. This network-based approach allows the tools of network analysis and graph theory to be brought to bear to analyze the overall topology of the decision network and to identify the key decision nodes and domains of outcomes associated with an adversary's decisionmaking. Analytical methods, such as centrality measures, can be used, for example, to identify the decision nodes that will have the greatest impact if they are disrupted. Community detection algorithms could be used to classify the different subdomains of the decision tree to allow modeling of the different likelihoods of operational success within different portions of the decision flow. The goal of framing the decision process in these terms is to provide a robust methodology by which the impacts of different deterrence strategies and intervention measures can be modeled and compared.

References

Air Force Doctrine Document 3-13, *Information Operations*, Washington, D.C.: U.S. Air Force, January 11, 2005, incorporating change 1, July 28, 2011. As of January 29, 2021:
https://fas.org/irp/doddir/usaf/afdd3-13.pdf

Bar-Yam, Yaneer, *Dynamics of Complex Systems*, Reading, Mass.: Perseus Books, 1997.

Bosse, Tibor, Alexei Sharpanskykh, and Jan Treur, "Modelling Complex Systems by Integration of Agent-Based and Dynamical Systems Models," in Ali Minai, Dan Braha, and Yaneer Bar-Yam, eds., *Unifying Themes in Complex Systems*, Berlin: Springer-Verlag Berlin, 2010, pp. 42–49.

Brinsmead, Thomas S., and Cliff Hooker, "Complex Systems Dynamics and Sustainability: Conception, Method and Policy," in Cliff Hooker, ed., *Philosophy of Complex Systems*, Vol. 10 (*Handbook of the Philosophy of Science*), Amsterdam, North Holland: Elsevier, 2011, pp. 809–838.

Capra, Fritjof, and Pier Luigi Luisi, *The Systems View of Life: A Unifying Vision*, Cambridge, UK: Cambridge University Press, 2016.

Choate, Jeffrey L., *Extending AFSIM with Behavioral Emergence*, master's thesis, Wright-Patterson Air Force Base, Ohio: Air Force Institute of Technology, March 23, 2017. As of January 17, 2021:
https://apps.dtic.mil/dtic/tr/fulltext/u2/1054622.pdf

Churchman, Charles West, "Guest Editorial: Wicked Problems," *Management Science*, Vol. 14, No. 4, December 1967, pp. B141–B142.

Davis, Paul K., "Strategic Planning Amidst Massive Uncertainty in Complex Adaptive Systems: the Case of Defense Planning," in Ali A. Minai and Yaneer Bar-Yam, *Unifying Themes in Complex Systems, New Research*, Volume IIIB, *Proceedings from the Third International Conference on Comple Systems*, Cambridge, Mass.: Springer, Berlin, 2006, pp. 201–214.

Davis, Paul K., Tim McDonald, Ann Pendleton-Jullian, Angela O'Mahony, and Osonde A. Osoba, "A Complex-Systems Agenda for Influencing Policy Studies," working paper, Santa Monica, Calif.: RAND Corporation, WR-1326, 2020. As of January 18, 2021:
https://www.rand.org/pubs/working_papers/WR1326.html

Diehl, Ernst, and John D. Sterman, "Effects of Feedback Complexity on Dynamic Decision Making," *Organizational Behavior and Human Decision Processes*, Vol. 62, No. 2, 1995, pp. 198–215.

Estrada, Ernesto, *The Structure of Complex Networks: Theory and Applications*, New York: Oxford Unviersity Press, 2011.

Garfinkel, Alan, Jane Shevtsov, and Yina Guo, *Modeling Life: The Mathematics of Biological Systems*, Cham, Switzerland: Springer International Publishing AG, 2017.

Grösser, Stefan N., "Complexity Management and System Dynamics Thinking," in Stefan N. Grösser, Arcadio Reyes-Lecuona, and Göran Granholm, eds., *Dynamics of Long-Life Assets*, Cham, Switzerland: Springer, 2017, pp. 69–92.

Henry, Ryan, Steven Berner, and David A. Shlapak, *Serious Analytical Gaming: The 360° Game for Multidimensional Analysis of Complex Problems*, Santa Monica, Calif.: RAND Corporation, RR-1764-OSD, 2017. As of January 18, 2021: https://www.rand.org/pubs/research_reports/RR1764.html

Hofstetter, Dominic, "Innovating in Complexity (Part II): From Single-Point Solutions to Directional Systems Innovation," *Medium*, July 25, 2019. As of January 18, 2021: https://medium.com/in-search-of-leverage/innovating-in-complexity-part-ii-from-single-point-solutions-to-directional-systems-innovation-dfb36fcfe50

Holland, John H., *Emergence: From Chaos to Order*, New York: Basic Books, 1999.

Holland, John H., *Complexity: A Very Short Introduction*, Oxford: Oxford University Press, 2014.

The Iraq Study Group, James A. Baker III, and Lee H. Hamilton, *The Iraq Study Group Report: The Way Forward—A New Approach*, New York: Vintage Books, December 2006.

Joint Publication 3-13.4, *Military Deception*, Washington, D.C.: U.S. Department of Defense, July 13, 2006. As of January 18, 2021: https://fas.org/irp/doddir/dod/jp3_13_4.pdf

Lawson, Sean T., *Nonlinear Science and Warfare: Chaos, Complexity and the U.S. Military in the Information Age*, New York: Routledge, 2013.

Lempert, Robert J., "A New Decision Sciences for Complex Systems," *Proceedings of the National Academy of Sciences*, Vol. 99, Supp. 3, May 2002, pp. 7309–7313.

Lempert, Robert J., Steven W. Popper, and Steven C. Bankes, *Shaping the Next One Hundred Years: New Methods for Quantitative, Long-Term Policy Analysis*, Santa Monica, Calif.: RAND Corporation, MR-1626-RPC, 2003. As of January 18, 2021: https://www.rand.org/pubs/monograph_reports/MR1626.html

Lempert, Robert J., Steven W. Popper, David G. Groves, Nidhi Kalra, Jordan R. Fischbach, Steven C. Bankes, Benjamin P. Bryant, Myles T. Collins, Klaus Keller, Andrew Hackbarth, et al., *Making Good Decisions Without Predictions: Robust Decision Making for Planning*

Under Deep Uncertainty, Santa Monica, Calif.: RAND Corporation, RB-9701, 2013. As of January 18, 2021:
https://www.rand.org/pubs/research_briefs/RB9701.html

Lingel, Sherrill, Matthew Sargent, Timothy R. Gulden, Tim McDonald, and Parousia Rockstroh, *Leveraging Complexity in Great-Power Competition and Warfare:* Volume I*, An Initial Exploration of How Complex Adaptive Systems Thinking Can Frame Opportunities and Challenges*, Santa Monica, Calif.: RAND Corporation, RR-A589-1, 2021. As of August 2021:
https://www.rand.org/pubs/research_reports/RRA589-1.html

Mazarr, Michael J., "Struggle in the Gray Zone and World Order," *War on the Rocks*, December 22, 2015. As of January 18, 2021:
https://warontherocks.com/2015/12/struggle-in-the-gray-zone-and-world-order/

Mazarr, Michael J., *Mastering the Gray Zone: Understanding a Changing Era of Conflict*, Carlisle Barracks, Pa.: United States Army War College Press, 2016.

Mazzocchi, Fulvio, "Complexity in Biology: Exceeding the Limits of Reductionism and Determinism Using Complexity Theory," *European Molecular Biology Organization*, Vol. 9, No. 1, January 2008, pp. 10–14.

Meadows, Donella, *Leverage Points: Places to Intervene in a System*, Hartland, Vt.: The Sustainability Institute, 1999. As of January 18, 2021:
http://donellameadows.org/wp-content/userfiles/Leverage_Points.pdf

Miller, John H., and Scott E. Page, *Complex Adaptive Systems: An Introduction to Computational Models of Social Life*, Princeton, N.J.: Princeton University Press, 2007.

Mitchell, Melanie, *Complexity: A Guided Tour*, Oxford: Oxford University Press, 2009.

Mitchell, Sandra D., *Unsimple Truths: Science, Complexity, and Policy*, Chicago: University of Chicago Press, 2009.

Monaghan, Sean, "Countering Hybrid Warfare: So What for the Joint Force?" *PRISM*, Vol. 8, No. 2, October 4, 2019, pp. 83–98.

Morris, Lyle J., Michael J. Mazarr, Jeffrey W. Hornung, Stephanie Pezard, Anika Binnendijk, and Marta Kepe, *Gaining Competitive Advantage in the Gray Zone: Response Options for Coercive Aggression Below the Threshold of Major War*, Santa Monica, Calif.: RAND Corporation, RR-2942-OSD, 2019. As of January 18, 2021:
https://www.rand.org/pubs/research_reports/RR2942.html

Murphy, Eric M., *Complex Adaptive Systems and the Development of Force Structures for the United States Air Force*, Drew Paper No. 18, Maxwell Air Force Base, Ala.: Air University Press, Air Force Research Institute, 2014. As of January 8, 2021:

https://www.airuniversity.af.edu/Portals/10/AUPress/Papers/DP_0018_MURPHY_COMPLE X_ADAPTIVE_SYSTEMS.PDF

Nason, Rick, *It's Not Complicated: The Art and Science of Complexity in Business*, Toronto: University of Toronto Press, 2017.

National Academies of Sciences, Engineering, and Medicine, *Guiding Cancer Control: A Path to Transformation*, Washington, D.C.: National Academies Press, 2019.

Petrovich, Daniel J., *Structural Emergence and the Collaborative Behavior of Autonomous Nano-Satellites*, thesis, Wright-Pattern Air Force Base, Ohio: Air Force Institute of Technology, March 8, 1999. As of January 18, 2021: https://apps.dtic.mil/dtic/tr/fulltext/u2/a361744.pdf

Pope, Charles, "Goldfein Stresses Promise of Multi-Domain Operations, Calls It 'the Single Most Critical' Tool for Winning Future High-End Fights," U.S. Air Force, July 18, 2019. As of January 18, 2021: https://www.af.mil/News/Article-Display/Article/1909269/goldfein-stresses-promise-of-multi-domain-operations-calls-it-the-single-most-c/

Richardson, Michael J., Alexandra Paxton, and Nikita Kuznetsov, "Nonlinear Methods for Understanding Complex Dynamical Phenomena in Psychological Science," *Psychological Science Agenda*, February 2017. As of January 8, 2021: https://www.apa.org/science/about/psa/2017/02/dynamical-phenomena

Rickles, Dean, Penelope Hawe, and Alan Shiell, "A Simple Guide to Chaos and Complexity," *Journal of Epidemiology Community Health*, Vol. 61, No. 11, November 2007, pp. 933–937.

Rittel, Horst W. J., and Melvin M. Webber, "Dilemmas in a General Theory of Planning," *Policy Sciences*, Vol. 4, 1973, pp. 155–169.

Ruiz-Martin, Cristina, Adolfo López Paredes, and Gabriel A. Wainer, "Applying Complex Network Theory to the Assessment of Organizational Resilience," *IFAC-PapersOnLine*, Vol. 48, No. 3, 2015, pp. 1224–1229.

Russell, Martha G., and Nataliya V. Smorodinskaya, "Leveraging Complexity for Ecosystemic Innovation," *Technological Forecasting and Social Change*, Vol. 136, November 2018, pp. 114–131.

Setear, John, *Simulating the Fog of War*, Santa Monica, Calif.: RAND Corporation, P-7511, 1989. As of January 18, 2021: https://www.rand.org/pubs/papers/P7511.html

Siegfried, Robert, *Modeling and Simulation of Complex Systems: A Framework for Efficient Agent-Based Modeling and Simulation*, Wiesbaden, Germany: Springer Vieweg, 2014.

Simon, Herbert A., "Theories of Bounded Rationality," in C. B. McGuire and Roy Radner, eds., *Decision and Organization*, Amsterdam: North-Holland Publishing Company, 1972, pp. 161–176.

Simon, Herbert A., *The Sciences of the Artificial*, 3rd ed., Cambridge, Mass.: MIT Press, 1996.

Soni, Devin, "Introduction to Markov Chains," *Towards Data Science*, March 5, 2018. As of March 20, 2020:
https://towardsdatascience.com/introduction-to-markov-chains-50da3645a50d

Von Clausewitz, Carl, *On War*, ed. and trans. Michael Howard and Peter Paret, Princeton, N.J.: Princeton University Press, 1989 [1832].

Zimmerman, S. Rebecca, Kimberly Jackson, Natasha Lander, Colin Roberts, Dan Madden, and Rebeca Orrie, *Movement and Maneuver: Culture and the Competition for Influence Among the U.S. Military Services*, Santa Monica, Calif.: RAND Corporation, RR-2270-OSD, 2019. As of January 18, 2021:
https://www.rand.org/pubs/research_reports/RR2270.html